I0759894

THE

GREAT NORTHERN

Expedition

FOREWORD

In 1725, Peter the Great of Russia commissioned Captain Vitus Bering to sail from the easternmost province of Kamchatka to the north, in hopes of finding a link between Asia and America. This expedition began by transporting tools, anchors, chains, tar, sails, and other supplies across roughly 5,600 miles (9,000 km) of Siberian wilderness. They then sailed a remote and inhospitable sea. Blinded by fog, Bering crossed the strait later named after him without sighting America and continued until the threat of the impending winter forced them to return. The mission was considered a failure.

Bering did not want to give up. So the Great Northern Expedition was born. It was to be a scientific venture of incredible scope and importance—it also promised to be a lengthy exile for its participants, so many officers brought their families, such as Bering and his lieutenant, the Swede Sven Waxell. In 1733, various convoys started leaving St. Petersburg. While his men drove the transports, Bering wrangled with Siberian administrative agents who were far from their superiors in the capital and much less willing to provide him with the assistance he had been promised. Bering was 60 years old when he finally sailed for America in 1741. Although the ships' logs note only the essentials, some of those who returned in 1742 recounted their memories of this incredible nine-year expedition.

My father had told me this expedition would be the most important thing we ever did. We were to map the Arctic coast of Siberia as well as parts of the North American coastline. Some members of the expedition were also expected to find a northern sea route to Japan.

THE GREAT NORTHERN *Expedition*

MARIA CRISTINA PRITELLI

My father is Sven Waxell, second-in-command to Captain Vitus Bering.

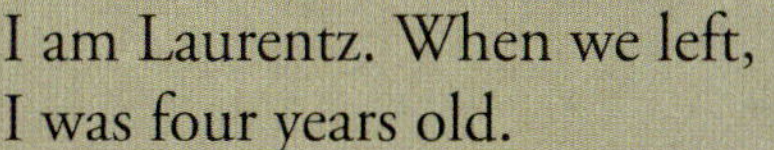

CREATIVE EDITIONS

Saint Petersburg, Russia
February 1733

My earliest memories of the journey
are ones of cold, of frosty air turning
words into crystals.

Then, frozen rivers melted into plains as wide as the sea, and the grass was clouded by mosquitoes.

But as soon as I could smell autumn, it was over, replaced with winter's white tablecloth spread over the taiga.

Lena River

June 1735

The boatmen's song signaled spring: "Yo, heave ho! Yo, heave ho!"

With the thaw, though, came the tigers, lurking behind the leaves.

When we reached Yakutsk, the summer sun never seemed to set, while the winter sun never seemed to rise. Other ships drifted down the Lena—some never to return—as we waited for two long years, gathering supplies and assembling equipment.

1737

Icy rivers and treacherous mountains made constant foes as we plodded toward Okhotsk, where we said good-bye to most of our family members.

The abundant salmon gave us hope in that otherwise abandoned port. From there, one ship left for Japan, and ours departed for Kamchatka.

At last, we reached the jumping-off point for our sea voyage! However, it would be another year before all the supplies could be gathered and we set sail on the newly built *St. Peter*, with the *St. Paul* as our companion.

Mount Saint Elias, Alaska

July 16, 1741

Soon separated from the *St. Paul* in a fog, what we did not realize then was that the map guiding us was largely a fiction. After wasting many days, supplies, and precious water, we spotted a mountain.

We made brief landfall on an island to replenish our water stores. One of the scientific assistants insisted on taking time to explore.

After almost three months at sea, with little water and stale rations, the dreaded scurvy took hold of our ship. And then we saw people!

We could not communicate with words, but my father sketched a kayaker as he approached the ship.

November 1741

Terrible storms spat those of us who were still alive out on an island. With winter coming, we dug holes in the sand and tried to fend off the foxes that haunted our days.

Death took Captain Bering in December and remained a constant and unwelcome companion until January 1742.

My father, now commander, led us through despair to survival. In spring, the sea was filled with food. We rebuilt a little *St. Peter* from the remains of the first ship.

We returned to Kamchatka with our heads full of tales and our hearts full of resolve. This ending is only the beginning.

Selected Bibliography

Frost, O. W. *Bering: The Russian Discovery of America*. New Haven: Yale University Press, 2003.

Lauridsen, Peter. *Vitus Bering, the Discoverer of Bering Strait*. Trans. Julius E. Olson. Cambridge: Cambridge University Press, 2012.

Steller, Georg Wilhelm. *Journal of a Voyage with Bering, 1741–1742*. Stanford: Stanford University Press, 1988.

Waxell, Sven. *The American Expedition*. London: William Hodge, 1952.

Edited by Kate Riggs, Designed by Rita Marshall
Published in 2026 by Creative Editions
P.O. Box 227, Mankato, MN 56002 USA
Creative Editions is an imprint of The Creative Company
www.thecreativecompany.us

Library of Congress Cataloging-in-Publication Data
Names: Pritelli, Maria Cristina author illustrator
Title: The Great Northern Expedition / [written and illustrated] by: Maria Cristina Pritelli. Description: Mankato, MN : Creative Editions, 2026. | Audience: Ages 8-12 | Audience: Grades 2-3 | Summary: "The Great Northern Expedition to Alaska, led by Russia in the 1700s, is reimagined through the eyes of Sven Waxell's son, capturing the beauty and ambition of the doomed voyage"— Provided by publisher. Identifiers: LCCN 2025015392 (print) | LCCN 2025015393 (ebook) | ISBN 9781568464015 hardcover | ISBN 9781568464022 ebook | Subjects: LCSH: Waxell, Laurentz, 1729–1781—Juvenile literature | Waxell, Sven Larsson, 1701–1761 or 1762—Juvenile literature | Kamchatskaiâ ėkspeditŝiiâ (2nd : 1733-1744) | Explorers—Russia—Biography—Juvenile literature / Classification: LCC G295 .P75 2026 (print) | LCC G295 (ebook) | DDC 917.9804/10922—dc23/eng/20250530
LC record available at https://lccn.loc.gov/2025015392
LC ebook record available at https://lccn.loc.gov/2025015393
First edition 9 8 7 6 5 4 3 2 1   Printed in China